*f*P

Also by Sam Harris

The End of Faith

Letter to a Christian Nation

The Moral Landscape

Lying

FREE WILL

SAM HARRIS

FREE PRESS

New York London Toronto Sydney New Delhi

FREE PRESS
A Division of Simon & Schuster, Inc.
1230 Avenue of the Americas
New York, NY 10020

First Free Press trade paperback edition March 2012

FREE PRESS and colophon are trademarks of Simon & Schuster, Inc.

For information about special discounts for bulk purchases, please contact Simon & Schuster Special Sales at 1-866-506-1949 or business@simonandschuster.com.

The Simon & Schuster Speakers Bureau can bring authors to your live event. For more information or to book an event contact the Simon & Schuster Speakers Bureau at 1-866-248-3049 or visit our website at www.simonspeakers.com.

Manufactured in the United States of America

3 5 7 9 10 8 6 4

Library of Congress Cataloging-in-Publication Data
Harris, Sam,
Free will / Sam Harris.—1st Free Press trade pbk. ed.
p. cm.
1. Free will and determinism. I. Title.
BJ1461.H2785 2012
123'.5—dc23 2011052177

ISBN 978-1-4516-8340-0
ISBN 978-1-4516-8347-9 (ebook)

Some of the material in this work was previously published in the author's book *The Moral Landscape*.

For Hitch

FREE WILL

The question of free will touches nearly everything we care about. Morality, law, politics, religion, public policy, intimate relationships, feelings of guilt and personal accomplishment—most of what is distinctly *human* about our lives seems to depend upon our viewing one another as autonomous persons, capable of free choice. If the scientific community were to declare free will an illusion, it would precipitate a culture war far more belligerent than the one that has been waged on the subject of evolution. Without free will, sinners and criminals would be nothing more than poorly calibrated clockwork, and any conception of justice that emphasized punishing them (rather than deterring, rehabilitating, or merely containing them) would appear utterly incongruous. And those of us who work hard and follow the rules would not "deserve" our success in any deep sense. It is not an accident that most people find these conclusions abhorrent. The stakes are high.

In the early morning of July 23, 2007, Steven Hayes and Joshua Komisarjevsky, two career criminals, arrived at the home of Dr. William and Jennifer Petit in Cheshire, a quiet town in central Connecticut. They found Dr. Petit asleep on a sofa in the sunroom. According to his taped confession, Komisarjevsky stood over the sleeping man for some minutes, hesitating, before striking him in the head with a baseball bat. He claimed that his victim's screams then triggered something within him, and he bludgeoned Petit with all his strength until he fell silent.

The two then bound Petit's hands and feet and went upstairs to search the rest of the house. They discovered Jennifer Petit and her daughters—Hayley, 17, and Michaela, 11—still asleep. They woke all three and immediately tied them to their beds.

At 7:00 a.m., Hayes went to a gas station and bought four gallons of gasoline. At 9:30, he drove Jennifer Petit to her bank to withdraw $15,000 in cash. The conversation between Jennifer and the bank teller suggests that she was unaware of her husband's injuries and believed that her captors would release her family unharmed.

While Hayes and the girls' mother were away, Komisarjevsky amused himself by taking naked photos of Michaela with his cell phone and masturbating on her. When Hayes returned with Jennifer, the two men divided up the money and briefly considered what they should do. They decided that Hayes should take Jennifer into the living room and rape her—which he did. He then strangled her, to the apparent surprise of his partner.

At this point, the two men noticed that William Petit had slipped his bonds and escaped. They began to panic. They quickly doused the house with gasoline and set it on fire. When asked by the police why he hadn't untied the two girls from their beds before lighting the blaze, Komisarjevsky said, "It just didn't cross my mind." The girls died of smoke inhalation. William Petit was the only survivor of the attack.

Upon hearing about crimes of this kind, most of us naturally feel that men like Hayes and Komisarjevsky should be held morally responsible for their actions. Had we been close to the Petit family, many of us would feel entirely justified in killing these monsters with our own hands. Do we care that Hayes has since shown signs of remorse and has attempted suicide? Not really. What about the fact that Komisarjevsky was repeatedly raped as a child? According to his journals, for as long as he can remember, he has known that he was "different" from other people, psychologically damaged, and capable of great coldness. He also claims to have been stunned by his own behavior in the Petit home: He was a career burglar, not a murderer, and he had not consciously intended to kill anyone. Such details might begin to give us pause.

As we will see, whether criminals like Hayes and Komisarjevsky can be trusted to honestly report their feelings and intentions is not the point: Whatever their conscious motives, these men cannot know why they are as they are. Nor can we account for why we are not like them. As sickening as I find their behavior, I have to admit that if I were to trade places with one of these men, atom for atom, I would *be* him: There is no extra part of me that could decide to see the world differently or to resist the impulse to victimize other people. Even if you believe that every human being harbors an immortal soul, the problem of responsibility remains: I cannot take credit for the fact that I do not have the soul of a psychopath. If I had truly been in Komisarjevsky's shoes on July 23, 2007—that is, if I had his genes and life experience and an identical brain (or soul) in an identical state—I would have acted exactly as he did. There is simply no intellectually respectable position from which to deny this. The role of luck, therefore, appears decisive.

Of course, if we learned that both these men had been suffering from brain tumors that explained their violent behavior, our moral intuitions would shift dramatically. But a neurological disorder appears to be just a special case of physical events giving rise to thoughts and actions. Understanding the neurophysiology of the brain, therefore, would seem to be as exculpatory as finding a tumor in it. How can we make sense of our lives, and hold people accountable for their choices, given the unconscious origins of our conscious minds?

Free will *is* an illusion. Our wills are simply not of our own making. Thoughts and intentions emerge from background causes of which we are unaware and over which we exert no conscious control. We do not have the freedom we think we have.

Free will is actually more than an illusion (or less), in that it cannot be made conceptually coherent. Either our wills are determined by prior causes and we are not responsible for them, or they are the product of chance and we are not responsible for them. If a man's choice to shoot the president is determined by a certain pattern of neural activity, which is in turn the product of prior causes—perhaps an unfortunate coincidence of bad genes, an unhappy childhood, lost sleep, and cosmic-ray bombardment—what can it possibly mean to say that his will is "free"? No one has ever described a way in which mental and physical processes could arise that would attest to the existence of such freedom. Most illusions are made of sterner stuff than this.

The popular conception of free will seems to rest on two assumptions: (1) that each of us could have behaved differently than we did in the past, and (2) that we are the conscious source of most of our thoughts and actions in the present. As we are about to see, however, both of these assumptions are false.

But the deeper truth is that free will doesn't even correspond to any *subjective* fact about us—and introspection soon proves as hostile to the idea as the laws of physics are. Seeming acts of volition merely arise spontaneously (whether caused, uncaused, or probabilistically inclined, it makes no difference) and cannot be traced to a point of origin in our conscious minds. A moment or two of serious self-scrutiny, and you might observe that you no more decide the next thought you think than the next thought I write.

The Unconscious Origins of the Will

We are conscious of only a tiny fraction of the information that our brains process in each moment.[1] Although we continually notice changes in our experience—in thought, mood, perception, behavior, etc.—we are utterly unaware of the neurophysiological events that produce them. In fact, we can be very poor witnesses to experience itself. By merely glancing at your face or listening to your tone of voice, others are often more aware of your state of mind and motivations than you are.

I generally start each day with a cup of coffee or tea—sometimes two. This morning, it was coffee (two). Why not tea? I am in no position to know. I wanted coffee more than I wanted tea today, and I was free to have what I wanted. Did I consciously choose coffee over tea? No. The choice was made for me by events in my brain that I, as the conscious witness of my thoughts and actions, could not inspect or influence. Could I have "changed my mind" and switched to tea before the coffee drinker in me could get his bearings? Yes, but this impulse would also have been the product of unconscious causes. Why didn't it arise this morning? Why might it arise in the future? I cannot know. The intention to do one thing and not another does not originate in consciousness—rather, it *appears* in consciousness, as does any thought or impulse that might oppose it.

The physiologist Benjamin Libet famously used EEG to show that activity in the brain's motor cortex can be detected some 300 milliseconds before a person feels that he has decided to move.[2] Another lab extended this work using functional magnetic resonance imaging (fMRI): Subjects were asked to press one of two buttons while watching a "clock" composed of a random sequence of letters appearing on a screen. They reported which letter was visible at the moment they decided to press one button or the other. The experimenters found two brain regions that contained information about which button subjects would press a full *7 to 10 seconds* before the decision was consciously made.[3] More recently, direct recordings from the cortex showed that the activity of merely 256 neurons was sufficient to predict with 80 percent accuracy a person's decision to move 700 milliseconds before he became aware of it.[4]

These findings are difficult to reconcile with the sense that we are the conscious authors of our actions. One fact now seems indisputable: Some moments before you are aware of what you will do next—a time in which you subjectively appear to have complete freedom to behave however you please—your brain has already determined what you will do. You then become conscious of this "decision" and believe that you are in the process of making it.

The distinction between "higher" and "lower" systems in the brain offers no relief: I, as the conscious witness of my experience, no more initiate events in my prefrontal cortex than I cause my heart to beat. There will always be some delay between the first neurophysiological events that kindle my next conscious thought and the thought itself. And even if there weren't—even if all mental states were truly coincident with their underlying brain states—I cannot decide what I will next think or intend until a thought or intention arises. What will my next mental state be? I do not know—it just happens. Where is the freedom in that?

* * *

Imagine a perfect neuroimaging device that would allow us to detect and interpret the subtlest changes in brain function. You might spend an hour thinking and acting freely in the lab, only to discover that the scientists scanning your brain had been able to produce a complete record of what you would think and do some moments in advance of each event. For instance, exactly 10 minutes and 10 seconds into the experiment, you decided to pick up a magazine from a nearby table and begin reading, but the scanner log shows this mental state arising at 10 minutes and 6 seconds—and the experimenters even knew which magazine you would choose. You read for a while and then got bored and stopped; the experimenters knew you would stop a second before you did and could tell which sentence would be the last you read.

And so it would go with everything else: You tried to recall the name of the lead experimenter, but you forgot it; a minute later you remembered it as "Brent" when it was actually "Brett." Next, you decided to go shopping for new shoes after you left the lab—but on second thought, you realized that your son would be getting out of school early that day, so you wouldn't have enough time to go shopping after all. Imagine what it would be like to see the time log of these mental events, alongside video of your associated behavior, demonstrating that the experimenters knew what you would think and do just before you did. You would, of course, continue to feel free in every present moment, but the fact that someone else could report what you were about to think and do would expose this feeling for what it is: an *illusion*. If the laws of nature do not strike most of us as incompatible with free will, that is because we have not imagined how human behavior would appear if all cause-and-effect relationships were understood.

It is important to recognize that the case I am building against free will does not depend upon philosophical materialism (the assumption that reality is, at bottom, purely physical). There is no question that (most, if not all) mental events are the product of physical events. The brain is a physical system, entirely beholden to the laws of nature—and there is every reason to believe that changes in its functional state and material structure entirely dictate our thoughts and actions. But even if the human mind were made of soul-stuff, nothing about my argument would change. The unconscious operations of a soul would grant you no more freedom than the unconscious physiology of your brain does.

If you don't know what your soul is going to do next, you are not in control. This is obviously true in all cases where a person wishes he could feel or behave differently than he does: Think of the millions of committed Christians whose souls happen to be gay, prone to obesity, or bored by prayer. However, free will is no more evident when a person does exactly what, in retrospect, he wishes he had done. The soul that allows you to stay on your diet is just as mysterious as the one that tempts you to eat cherry pie for breakfast.

There is a distinction between voluntary and involuntary actions, of course, but it does nothing to support the common idea of free will (nor does it depend upon it). A voluntary action is accompanied by the felt intention to carry it out, whereas an involuntary action isn't. Needless to say, this difference is reflected at the level of the brain. And what a person consciously intends to do says a lot about him. It makes sense to treat a man who enjoys murdering children differently from one who accidentally hit and killed a child with his car—because the conscious intentions of the former give us a lot of information about how he is likely to behave in the future. But where intentions themselves come from, and what determines their character in every instance, remains perfectly mysterious in subjective terms. Our sense of free will results from a failure to appreciate this: We do not know what we intend to do until the intention itself arises. To understand this is to realize that we are not the authors of our thoughts and actions in the way that people generally suppose.

Of course, this insight does not make social and political freedom any less important. The freedom to do what one intends, and not to do otherwise, is no less valuable than it ever was. Having a gun to your head is still a problem worth rectifying, wherever intentions come from. But the idea that we, as conscious beings, are deeply responsible for the character of our mental lives and subsequent behavior is simply impossible to map onto reality.

Consider what it would take to actually have free will. You would need to be aware of all the factors that determine your thoughts and actions, and you would need to have complete control over those factors. But there is a paradox here that vitiates the very notion of freedom—for what would influence the influences? More influences? None of these adventitious mental states are the real you. You are not controlling the storm, and you are not lost in it. You *are* the storm.

Changing the Subject

It is safe to say that no one was ever moved to entertain the existence of free will because it holds great promise as an abstract idea. The endurance of this notion is attributable to the fact that most of us *feel* that we freely author our own thoughts and actions (however difficult it may be to make sense of this in logical or scientific terms). Thus the idea of free will emerges from a felt experience. It is, however, very easy to lose sight of this psychological truth once we begin talking philosophy.

In the philosophical literature, one finds three main approaches to the problem: *determinism*, *libertarianism*, and *compatibilism*. Both determinism and libertarianism hold that if our behavior is fully determined by background causes, free will is an illusion. (For this reason they are both referred to as "incompatibilist" views.) Determinists believe that we live in such a world, while libertarians (no relation to the political philosophy that goes by this name) imagine that human agency must magically rise above the plane of physical causation. Libertarians sometimes invoke a metaphysical entity, such as a soul, as the vehicle for our freely acting wills. Compatibilists, however, claim that determinists and libertarians are both confused and that free will is compatible with the truth of determinism.

Today, the only philosophically respectable way to endorse free will is to be a compatibilist—because we know that determinism, in every sense relevant to human behavior, is true. Unconscious neural events determine our thoughts and actions—and are themselves determined by prior causes of which we are subjectively unaware. However, the "free will" that compatibilists defend is not the free will that most people feel they have.

Compatibilists generally claim that a person is free as long as he is free from any outer or inner compulsions that would prevent him from acting on his actual desires and intentions. If you want a second scoop of ice cream and no one is forcing you to eat it, then eating a second scoop is fully demonstrative of your freedom of will. The truth, however, is that people claim greater autonomy than this. Our moral intuitions and sense of personal agency are anchored to a felt sense that we are the *conscious source* of our thoughts and actions. When deciding whom to marry or which book to read, we do not feel compelled by prior events over which we have no control. The freedom that we presume for ourselves and readily attribute to others is felt to slip the influence of impersonal background causes. And the moment we see that such causes are fully effective—as any detailed account of the neurophysiology of human thought and behavior would reveal—we can no longer locate a plausible hook upon which to hang our conventional notions of personal responsibility.[5]

What does it mean to say that rapists and murderers commit their crimes of their own free will? If this statement means anything, it must be that they could have behaved differently—not on the basis of random influences over which they have no control, but because they, as conscious agents, were free to think and act in other ways. To say that they were free *not* to rape and murder is to say that they could have resisted the impulse to do so (or could have avoided feeling such an impulse altogether)—with the universe, including their brains, in precisely the same state it was in at the moment they committed their crimes. Assuming that violent criminals have such freedom, we reflexively blame them for their actions. But without it, the place for our blame suddenly vanishes, and even the most terrifying sociopaths begin to seem like victims themselves. The moment we catch sight of the stream of causes that precede their conscious decisions, reaching back into childhood and beyond, their culpability begins to disappear.

Compatibilists have produced a vast literature in an effort to finesse this problem.[6] More than in any other area of academic philosophy, the result resembles theology. (I suspect this is not an accident. The effort has been primarily one of not allowing the laws of nature to strip us of a cherished illusion.) According to compatibilists, if a man wants to commit murder, and does so because of this desire, his actions attest to his freedom of will. From both a moral and a scientific perspective, this seems deliberately obtuse. People have many competing desires—and some desires appear pathological (that is, *un*desirable) even to those in their grip. Most people are ruled by many mutually incompatible goals and aspirations: You want to finish your work, but you are also inclined to stop working so that you can play with your kids. You aspire to quit smoking, but you also crave another cigarette. You are struggling to save money, but you are also tempted to buy a new computer. Where is the freedom when one of these opposing desires inexplicably triumphs over its rival?

The problem for compatibilism runs deeper, however—for where is the freedom in wanting what one wants without any internal conflict whatsoever? Where is the freedom in being perfectly satisfied with your thoughts, intentions, and subsequent actions when they are the product of prior events that you had absolutely no hand in creating?

For instance, I just drank a glass of water and feel absolutely at peace with the decision to do so. I was thirsty, and drinking water is fully congruent with my vision of who I want to be when in need of a drink. Had I reached for a beer this early in the day, I might have felt guilty; but drinking a glass of water at any hour is blameless, and I am quite satisfied with myself. Where is the freedom in this? It may be true that if I had wanted to do otherwise, I would have, but I am nevertheless compelled to do what I effectively want. And I cannot determine my wants, or decide which will be effective, in advance. My mental life is simply given to me by the cosmos. Why didn't I decide to drink a glass of juice? The thought never occurred to me. Am I free to do *that which does not occur to me to do*? Of course not.

And there is no way I can influence my desires—for what tools of influence would I use? Other desires? To say that I would have done otherwise had I wanted to is simply to say that I would have lived in a different universe had I been in a different universe. Compatibilism amounts to nothing more than an assertion of the following creed: *A puppet is free as long as he loves his strings.*

Compatibilists like my friend Daniel Dennett[7] insist that even if our thoughts and actions are the product of unconscious causes, they are still *our* thoughts and actions. Anything that our brains do or decide, whether consciously or not, is something that *we* have done or decided. The fact that we cannot always be subjectively aware of the causes of our actions does not negate free will—because our unconscious neurophysiology is just as much "us" as our conscious thoughts are. Consider the following, from Tom Clark of the Center for Naturalism:

> Harris is of course right that we don't have conscious access to the neurophysiological processes that underlie our choices. But, as Dennett often

points out, these processes are as much our own, just as much part of who we are as persons, just as much *us*, as our conscious awareness. We shouldn't alienate ourselves from our own neurophysiology and suppose that the conscious self, what Harris thinks of as constituting the *real* self (and as many others do, too, perhaps), is being pushed around at the mercy of our neurons. Rather, as identifiable individuals we consist (among other things) of neural processes, some of which support consciousness, some of which don't. So it isn't an illusion, as Harris says, that we are authors of our thoughts and actions; we are not mere witnesses to what causation cooks up. We as physically instantiated persons really do deliberate and choose and act, even if consciousness isn't ultimately in charge. So the feeling of authorship and control is veridical.

Moreover, the neural processes that (somehow—the hard problem of consciousness) support consciousness *are* essential to choosing, since the evidence strongly suggests they are associated with flexible action and information integration in service to behavior control. But it's doubtful that consciousness (phenomenal experience) per se adds

> anything to those neural processes in controlling action.
>
> It's true that human persons don't have contra-causal free will. We are not self-caused little gods. But we are just as real as the genetic and environmental processes which created us and the situations in which we make choices. The deliberative machinery supporting effective action is just as real and causally effective as any other process in nature. So we don't have to talk *as if* we are real agents in order to concoct a motivationally useful *illusion* of agency, which is what Harris seems to recommend we do near the end of his remarks on free will. Agenthood survives determinism, no problem.[8]

This perfectly articulates the difference between Dennett's view and my own (Dennett agrees[9]). As I have said, I think compatibilists like Dennett change the subject: They trade a psychological fact—the subjective experience of being a conscious agent—for a conceptual understanding of ourselves as persons. This is a bait and switch. The psychological truth is that people feel identical to a certain channel of information in their conscious minds. Dennett is simply asserting that we are more than this—we are coterminous with everything that goes on inside our bodies, whether we are conscious of it or not. This is like saying we are made of stardust—which we are. But we don't *feel* like stardust. And the knowledge that we are stardust is not driving our moral intuitions or our system of criminal justice.[10]

At this moment, you are making countless unconscious "decisions" with organs other than your brain—but these are not events for which you feel responsible. Are *you* producing red blood cells and digestive enzymes at this moment? Your body is doing these things, of course, but if it "decided" to do otherwise, you would be the victim of these changes, rather than their cause. To say that you are responsible for everything that goes on inside your skin because it's all "you" is to make a claim that bears absolutely no relationship to the feelings of agency and moral responsibility that have made the idea of free will an enduring problem for philosophy.

There are more bacteria in your body than there are human cells. In fact, 90 percent of the cells in your body are microbes like *E. coli* (and 99 percent of the functional genes in your body belong to them). Many of these organisms perform necessary functions—they are "you" in some wider sense. Do you feel identical to them? If they misbehave, are you morally responsible?

People feel (or presume) an authorship of their thoughts and actions that is illusory. If we were to detect their conscious choices on a brain scanner seconds before they were aware of them, they would be rightly astonished—because this would directly challenge their status as conscious agents in control of their inner lives. We know that we could perform such an experiment, at least in principle, and if we tuned the machine correctly, subjects would feel that we were reading their minds (or controlling them).[11]

We know, in fact, that we sometimes feel responsible for events over which we have no causal influence. Given the right experimental manipulations, people can be led to believe that they consciously intended an action when they neither chose it nor had control over their movements. In one experiment, subjects were asked to select pictures on a screen using a computer's cursor. They tended to believe that they had intentionally guided the cursor to a specific image even when it was under the full control of another person, as long as they heard the name of the image just before the cursor stopped.[12] People who are susceptible to hypnosis can be given elaborate suggestions to perform odd tasks, and when asked why they have done these things, many will confabulate—giving reasons for their behavior that have nothing to do with its actual cause. There is no question that our attribution of agency can be gravely in error. I am arguing that it always is.

Imagine that a person claims to have no need to eat food of any kind—rather, he can live on light. From time to time, an Indian yogi will make such a boast, much to the merriment of skeptics. Needless to say, there is no reason to take such claims seriously, no matter how thin the yogi. However, a compatibilist like Dennett could come to the charlatan's defense: The man *does* live on light—we all do—because when you trace the origin of any food, you arrive at something that depends on photosynthesis. By eating beef, we consume the grass the cow ate, and the grass ate sunlight. So the yogi is no liar after all. But that's not the ability the yogi was advertising, and his actual claim remains dishonest (or delusional). This is the trouble with compatibilism. It solves the problem of "free will" by ignoring it.

How can we be "free" as conscious agents if everything that we consciously intend is caused by events in our brain that we *do not* intend and of which we are entirely unaware? We can't. To say that "my brain" decided to think or act in a particular way, whether consciously or not, and that this is the basis for my freedom, is to ignore the very source of our belief in free will: the feeling of *conscious* agency. People *feel* that they are the authors of their thoughts and actions, and this is the only reason why there seems to be a problem of free will worth talking about.

Cause and Effect

In physical terms, we know that every human action can be reduced to a series of impersonal events: Genes are transcribed, neurotransmitters bind to their receptors, muscle fibers contract, and John Doe pulls the trigger on his gun. But for our commonsense notions of human agency and morality to hold, it seems that our actions cannot be merely lawful products of our biology, our conditioning, or anything else that might lead others to predict them. Consequently, some scientists and philosophers hope that chance or quantum uncertainty can make room for free will.

For instance, the biologist Martin Heisenberg has observed that certain processes in the brain, such as the opening and closing of ion channels and the release of synaptic vesicles, occur at random, and cannot therefore be determined by environmental stimuli. Thus, much of our behavior can be considered truly "self-generated"—and therein, he imagines, lies a basis for human freedom. But how do events of this kind justify the feeling of free will? "Self-generated" in this sense means only that certain events originate in the brain.

If my decision to have a second cup of coffee this morning was due to a random release of neurotransmitters, how could the indeterminacy of the initiating event count as the free exercise of my will? Chance occurrences are by definition ones for which I can claim no responsibility. And if certain of my behaviors are truly the result of chance, they should be surprising *even to me*. How would neurological ambushes of this kind make me free?

Imagine what your life would be like if *all* your actions, intentions, beliefs, and desires were randomly "self-generated" in this way. You would scarcely seem to have a mind at all. You would live as one blown about by an internal wind. Actions, intentions, beliefs, and desires can exist only in a system that is significantly constrained by patterns of behavior and the laws of stimulus-response. The possibility of reasoning with other human beings—or, indeed, of finding their behaviors and utterances comprehensible at all—depends on the assumption that their thoughts and actions will obediently ride the rails of a shared reality. This is true as well when attempting to understand one's own behavior. In the limit, Heisenberg's "self-generated" mental events would preclude the existence of any mind at all.

The indeterminacy specific to quantum mechanics offers no foothold: If my brain is a quantum computer, the brain of a fly is likely to be a quantum computer, too. Do flies enjoy free will? Quantum effects are unlikely to be biologically salient in any case. They play a role in evolution because cosmic rays and other high-energy particles cause point mutations in DNA (and the behavior of such particles passing through the nucleus of a cell is governed by the laws of quantum mechanics). Evolution, therefore, seems unpredictable in principle.[13] But few neuroscientists view the brain as a quantum computer. And even if it were, quantum indeterminacy does nothing to make the concept of free will scientifically intelligible. In the face of any real independence from prior events, every thought and action would seem to merit the statement "I don't know what came over me."

If determinism is true, the future is set—and this includes all our future states of mind and our subsequent behavior. And to the extent that the law of cause and effect is subject to indeterminism—quantum or otherwise—we can take no credit for what happens. There is no combination of these truths that seems compatible with the popular notion of free will.

Choices, Efforts, Intentions

When we consider human behavior, the difference between premeditated, voluntary action and mere accident seems immensely consequential. As we will see, this distinction can be preserved—and with it, our most important moral and legal concerns—while banishing the idea of free will once and for all.

Certain states of consciousness seem to arise automatically, beyond the sphere of our intentions. Others seem self-generated, deliberative, and subject to our will. When I hear the sound of a leaf blower outside my window, it merely impinges upon my consciousness: I haven't brought it into being, and I cannot stop it at will. I can try to put the sound out of my mind by focusing on something else—my writing, for instance—and this act of directing attention feels different from merely hearing a sound. I am *doing* it. Within certain limits, I seem to choose what I pay attention to. The sound of the leaf blower intrudes, but I can seize the spotlight of my attention in the next moment and aim it elsewhere. This difference between nonvolitional and volitional states of mind is reflected at the level of the brain—for they are governed by different systems. And the difference between them must, in part, produce the felt sense that there is a conscious self endowed with freedom of will.

As we have begun to see, however, this feeling of freedom arises from our moment-to-moment ignorance of the prior causes of our thoughts and actions. The phrase "free will" describes what it *feels* like to identify with certain mental states as they arise in consciousness. Thoughts like "What should I get my daughter for her birthday? I know—I'll take her to a pet store and have her pick out some tropical fish" convey the apparent reality of choices, freely made. But from a deeper perspective (speaking both objectively and subjectively), thoughts simply arise unauthored and yet author our actions.

This is not to say that conscious awareness and deliberative thinking serve no purpose. Indeed, much of our behavior depends on them. I might unconsciously shift in my seat, but I cannot unconsciously decide that the pain in my back warrants a trip to a physical therapist.

To do the latter, I must become aware of the pain and be consciously motivated to do something about it. Perhaps it would be possible to build an insentient robot capable of these states—but in our case, certain behaviors seem to require the presence of conscious thought. And we know that the brain systems that allow us to reflect upon our experience are different from those involved when we automatically react to stimuli. So consciousness, in this sense, is not inconsequential.[14] And yet the entire process of becoming aware of the pain in my back, thinking about it, and seeking a remedy for it results from processes of which I am completely unaware. Did I, the conscious person, create my pain? No. It simply appeared. Did I create the thoughts about it that led me to consider physical therapy? No. They, too, simply appeared. This process of conscious deliberation, while different from unconscious reflex, offers no foundation for freedom of will.

As Dan Dennett and many others have pointed out, people generally confuse determinism with fatalism. This gives rise to questions like "If everything is determined, why should I do anything? Why not just sit back and see what happens?" This is pure confusion. To sit back and see what happens is itself a choice that will produce its own consequences. It is also extremely difficult to do: Just try staying in bed all day waiting for something to happen; you will find yourself assailed by the impulse to get up and do something, which will require increasingly heroic efforts to resist.

And the fact that our choices depend on prior causes does not mean that they don't matter. If I had not decided to write this book, it wouldn't have written itself. My choice to write it was unquestionably the primary cause of its coming into being. Decisions, intentions, efforts, goals, willpower, etc., are causal states of the brain, leading to specific behaviors, and behaviors lead to outcomes in the world. Human choice, therefore, is as important as fanciers of free will believe. But the next choice you make will come out of the darkness of prior causes that you, the conscious witness of your experience, did not bring into being.

Therefore, while it is true to say that a person would have done otherwise if he had chosen to do otherwise, this does not deliver the kind of free will that most people seem to cherish—because a person's "choices" merely appear in his mind as though sprung from the void. From the perspective of your conscious awareness, you are no more responsible for the next thing you think (and therefore do) than you are for the fact that you were born into this world.

Let's say your life has gone off track. You used to be very motivated, inspired by your opportunities, and physically fit, but now you are lazy, easily discouraged, and overweight. How did you get this way? You might be able to tell a story about how your life unraveled, but you cannot truly account for why you let it happen. And now you want to escape this downward trend and change yourself through an act of will.

You begin reading self-help books. You change your diet and join a gym. You decide to go back to school. But after six months of effort, you are no closer to living the life you want than you were before. The books failed to make an impact on you; your diet and fitness regime proved impossible to maintain; and you got bored with school and quit. Why did you encounter so many obstacles in yourself? You have no idea. You tried to change your habits, but your habits appear to be stronger than you are. Most of us know what it is like to fail in this way—and these experiences are not even slightly suggestive of freedom of will.

But you woke up this morning feeling even greater resolve. Enough is enough! Now you have a will of steel. Before stepping out of bed you had a brilliant idea for a website—and the discovery that the domain name was available for only 10 dollars has filled you with confidence. You are now an entrepreneur! You share the idea with several smart people, and they think it is guaranteed to make you rich.

The wind is at your back, your sails are full, and you are tacking furiously. As it turns out, a friend of yours is also a close friend of Tim Ferriss, the famous lifestyle coach and fitness guru. Ferriss offers to consult with you about your approach to diet and exercise. You find this meeting extremely helpful—and afterward you discover a reservoir of discipline in yourself that you didn't know was there. Over the next four months you swap 20 pounds of fat for 20 pounds of muscle. You weigh the same, but you are fully transformed. Your friends can't believe what you have accomplished. Even your enemies begin to ask you for advice.

You feel entirely different about your life, and the role that discipline, choice, and effort have played in your resurrection cannot be denied. But how can you account for your ability to make these efforts today and not a year ago? Where did this idea for a website come from? It just appeared in your mind. Did *you*, as the conscious agent you feel yourself to be, *create* it? (If so, why not just create the next one right now?) How can you explain the effect that Tim Ferriss's advice had on you? How can you explain your ability to respond to it?

If you pay attention to your inner life, you will see that the emergence of choices, efforts, and intentions is a fundamentally mysterious process. Yes, you can decide to go on a diet—and we know a lot about the variables that will enable you to stick to it—but you cannot know why you were finally able to adhere to this discipline when all your previous attempts failed. You might have a story to tell about why things were different this time around, but it would be nothing more than a post hoc description of events that you did not control. Yes, you can do what you want—but you cannot account for the fact that your wants are effective in one case and not in another (and you certainly can't choose your wants in advance). You wanted to lose weight for years. Then you *really* wanted to. What's the difference? Whatever it is, it's not a difference that *you* brought into being.

You are not in control of your mind—because you, as a conscious agent, are only *part* of your mind, living at the mercy of other parts.[15] You can do what you decide to do—but you cannot decide what you will decide to do. Of course, you can create a framework in which certain decisions are more likely than others—you can, for instance, purge your house of all sweets, making it very unlikely that you will eat dessert later in the evening—but you cannot know why you were able to submit to such a framework today when you weren't yesterday.

So it's not that willpower isn't important or that it is destined to be undermined by biology. Willpower is itself a biological phenomenon. You can change your life, and yourself, through effort and discipline—but you have whatever capacity for effort and discipline you have in this moment, and not a scintilla more (or less). You are either lucky in this department or you aren't—and you cannot make your own luck.

Many people believe that human freedom consists in our ability to do what, upon reflection, we believe we should do—which often means overcoming our short-term desires and following our long-term goals or better judgment. This is certainly an ability that people possess, to a greater or lesser degree, and which other animals appear to lack, but it is nevertheless a capacity of our minds that has unconscious roots.

You have not built your mind. And in moments in which you *seem* to build it—when you make an effort to change yourself, to acquire knowledge, or to perfect a skill—the only tools at your disposal are those that you have inherited from moments past.

Choices, efforts, intentions, and reasoning influence our behavior—but they are themselves part of a chain of causes that precede conscious awareness and over which we exert no ultimate control. My choices matter—and there are paths toward making wiser ones—but I cannot choose what I choose. And if it ever appears that I do—for instance, after going back and forth between two options—I do not *choose* to choose what I choose. There is a regress here that always ends in darkness. I must take a first step, or a last one, for reasons that are bound to remain inscrutable.[16]

Many people believe that this problem of regress is a false one. Certain compatibilists insist that freedom of will is synonymous with the idea that one could have thought or acted differently. However, to say that I could have done otherwise is merely to think the thought "I could have done otherwise" after doing whatever I in fact did. This is an empty affirmation.[17] It confuses hope for the future with an honest account of the past. What I will do next, and why, remains, at bottom, a mystery—one that is fully determined by the prior state of the universe and the laws of nature (including the contributions of chance). To declare my "freedom" is tantamount to saying, "I don't know why I did it, but it's the sort of thing I tend to do, and I don't mind doing it."

One of the most refreshing ideas to come out of existentialism (perhaps the only one) is that we are free to interpret and reinterpret the meaning of our lives. You can consider your first marriage, which ended in divorce, to be a "failure," or you can view it as a circumstance that caused you to grow in ways that were crucial to your future happiness. Does this freedom of interpretation require free will? No. It simply suggests that different ways of thinking have different consequences. Some thoughts are depressing and disempowering; others inspire us. We can pursue any line of thought we want—but our choice is the product of prior events that we did not bring into being.

Take a moment to think about the context in which

your next decision will occur: You did not pick your parents or the time and place of your birth. You didn't choose your gender or most of your life experiences. You had no control whatsoever over your genome or the development of your brain. And now your brain is making choices on the basis of preferences and beliefs that have been hammered into it over a lifetime—by your genes, your physical development since the moment you were conceived, and the interactions you have had with other people, events, and ideas. Where is the freedom in this? Yes, you are free to do what you want even now. But where did your desires come from?

Writing for *The New York Times*, the philosopher Eddy Nahmias criticized arguments of the sort I have presented here:

> Many philosophers, including me, understand free will as a set of capacities for imagining future courses of action, deliberating about one's reasons for choosing them, planning one's actions in light of this deliberation and controlling actions in the face of competing desires. We act of our own free

> will to the extent that we have the opportunity to exercise these capacities, without unreasonable external or internal pressure. We are responsible for our actions roughly to the extent that we possess these capacities and we have opportunities to exercise them.[18]

There is no question that human beings can imagine and plan for the future, weigh competing desires, etc.—and that losing these capacities would greatly diminish us. External and internal pressures of various kinds can be present or absent while a person imagines, plans, and acts—and such pressures determine our sense of whether he is morally responsible for his behavior. However, these phenomena have nothing to do with free will.

For instance, in my teens and early twenties I was a devoted student of the martial arts. I practiced incessantly and taught classes in college. Recently, I began training again, after a hiatus of more than 20 years. Both the cessation and the renewal of my interest in martial arts seem to be pure expressions of the freedom that Nahmias attributes to me. I have been under no "unreasonable external or internal pressure." I have done exactly what I wanted to do. I wanted to stop training, and I stopped. I wanted to start again, and now I train several times a week. All this has been associated with conscious thought and acts of apparent self-control.

However, when I look for the psychological cause of my behavior, I find it utterly mysterious. Why did I stop training 20 years ago? Well, certain things just became more important to me. But *why* did they become more important to me—and why precisely then and to that degree? And why did my interest in martial arts suddenly reemerge after decades of hibernation? I can consciously weigh the effects of certain influences—for instance, I recently read Rory Miller's excellent book *Meditations on Violence.* But why did I read this book? I have no idea. And why did I find it compelling? And why was it sufficient to provoke action on my part (if, indeed, it was the proximate cause of my behavior)? And why so much action? I'm now practicing two martial arts and also training with Miller and other self-defense experts. What in hell is going on here? Of course, I could tell a story about why I'm doing what I'm doing—which would amount to my telling you why I think such training is a good idea, why I enjoy it, etc.—but the actual explanation for my behavior is hidden from me. And it is perfectly obvious that I, as the conscious witness of my experience, am not the deep cause of it.

After reading the previous paragraph, some of you will think, "That Miller book sounds interesting!" and you will buy it. Some will think no such thing. Of those who buy the book, some will find it extremely useful. Others might put it down without seeing the point. Others will place it on the shelf and forget to read it. Where is the freedom in any of this? You, as the conscious agent who reads these words, are in no position to determine which of these bins you might fall into. And if you decide to switch bins—"I wasn't going to buy the book, but now I will, just to spite you!"—you cannot account for that decision either. You will do whatever it is you do, and it is meaningless to assert that you could have done otherwise.

Might the Truth Be Bad for Us?

Many people worry that free will is a necessary illusion—and that without it we will fail to live creative and fulfilling lives. This concern isn't entirely unjustified. One study found that having subjects read an argument against the existence of free will made them more likely to cheat on a subsequent exam.[19] Another found such subjects to be less helpful and more aggressive.[20] It is surely conceivable that knowing (or emphasizing) certain truths about the human mind could have unfortunate psychological and/or cultural consequences. However, I'm not especially worried about degrading the morality of my readers by publishing this book.

Speaking from personal experience, I think that losing the sense of free will has only improved my ethics—by increasing my feelings of compassion and forgiveness, and diminishing my sense of entitlement to the fruits of my own good luck. Is such a state of mind always desirable? Probably not. If I were teaching a self-defense class for women, I would consider it quite counterproductive to emphasize that all human behavior, including a woman's response to physical attack, is determined by a prior state of the universe, and that all rapists are, at bottom, unlucky—being themselves victims of prior causes that they did not create. There are scientific, ethical, and practical truths appropriate to every occasion—and an injunction like "Just gouge the bastard's eyes" surely has its place. There is no contradiction here. Our interests in life are not always served by viewing people and things as collections of atoms—but this doesn't negate the truth or utility of physics.

Losing a belief in free will has not made me fatalistic—in fact, it has increased my feelings of freedom. My hopes, fears, and neuroses seem less personal and indelible. There is no telling how much I might change in the future. Just as one wouldn't draw a lasting conclusion about oneself on the basis of a brief experience of indigestion, one needn't do so on the basis of how one has thought or behaved for vast stretches of time in the past. A creative change of inputs to the system—learning new skills, forming new relationships, adopting new habits of attention—may radically transform one's life.

Becoming sensitive to the background causes of one's thoughts and feelings can—paradoxically—allow for greater creative control over one's life. It is one thing to bicker with your wife because you are in a bad mood; it is another to realize that your mood and behavior have been caused by low blood sugar. This understanding reveals you to be a biochemical puppet, of course, but it also allows you to grab hold of one of your strings: A bite of food may be all that your personality requires. Getting behind our conscious thoughts and feelings can allow us to steer a more intelligent course through our lives (while knowing, of course, that we are ultimately being steered).

Moral Responsibility

The belief in free will has given us both the religious conception of "sin" and our commitment to retributive justice. The U.S. Supreme Court has called free will a "universal and persistent" foundation for our system of law, distinct from "a deterministic view of human conduct that is inconsistent with the underlying precepts of our criminal justice system" (*United States* v. *Grayson*, 1978). Any intellectual developments that threatened free will would seem to put the ethics of punishing people for their bad behavior in question.

The great worry, of course, is that an honest discussion of the underlying causes of human behavior appears to leave no room for moral responsibility. If we view people as neuronal weather patterns, how can we coherently speak about right and wrong or good and evil? These notions seem to depend upon people being able to freely choose how to think and act. And if we remain committed to seeing people as *people*, we must find some notion of personal responsibility that fits the facts.

Happily, we can. What does it mean to take responsibility for an action? Yesterday I went to the market; I was fully clothed, did not steal anything, and did not buy anchovies. To say that I was responsible for my behavior is simply to say that what I did was sufficiently in keeping with my thoughts, intentions, beliefs, and desires to be considered an extension of them. If I had found myself standing in the market naked, intent upon stealing as many tins of anchovies as I could carry, my behavior would be totally out of character; I would feel that I was not in my right mind, or that I was otherwise not responsible for my actions. Judgments of responsibility depend upon the overall complexion of one's mind, not on the metaphysics of mental cause and effect.

Consider the following examples of human violence:

1. A four-year-old boy was playing with his father's gun and killed a young woman. The gun had been kept loaded and unsecured in a dresser drawer.
2. A 12-year-old boy who had been the victim of

continual physical and emotional abuse took his father's gun and intentionally shot and killed a young woman because she was teasing him.

3. A 25-year-old man who had been the victim of continual abuse as a child intentionally shot and killed his girlfriend because she left him for another man.
4. A 25-year-old man who had been raised by wonderful parents and never abused intentionally shot and killed a young woman he had never met "just for the fun of it."
5. A 25-year-old man who had been raised by wonderful parents and never abused intentionally shot and killed a young woman he had never met "just for the fun of it." An MRI of the man's brain revealed a tumor the size of a golf ball in his medial prefrontal cortex (a region responsible for the control of emotion and behavioral impulses).

In each case a young woman died, and in each case her death was the result of events arising in the brain of another human being. But the degree of moral outrage we feel depends on the background conditions described in each case. We suspect that a four-year-old child cannot truly kill someone on purpose and that the intentions of a 12-year-old do not run as deep as those of an adult. In cases 1 and 2, we know that the brain of the killer has not fully matured and that not all the responsibilities of personhood have yet been conferred. The history of abuse and the precipitating circumstance in case 3 seem to mitigate the man's guilt: This was a crime of passion committed by a person who had himself suffered at the hands of others. In 4 there has been no abuse, and the motive brands the perpetrator a psychopath. Case 5 involves the same psychopathic behavior and motive, but a brain tumor somehow changes the moral calculus entirely: Given its location, it seems to divest the killer of all responsibility for his crime. And it works this miracle even if the man's subjective experience was identical to that of the psychopath in case 4—for the moment we understand that his feelings had a physical cause, a brain tumor, we cannot help seeing him as a victim of his own biology.

How can we make sense of these gradations of moral responsibility when brains and their background influences are in every case, and to exactly the same degree, the real cause of a woman's death?

We need not have any illusions that a causal agent lives within the human mind to recognize that certain people are dangerous. What we condemn most in another person is *the conscious intention to do harm*. Degrees of guilt can still be judged by reference to the facts of a case: the personality of the accused, his prior offenses, his patterns of association with others, his use of intoxicants, his confessed motives with regard to the victim, etc. If a person's actions seem to have been entirely out of character, this might influence our view of the risk he now poses to others. If the accused appears unrepentant and eager to kill again, we need entertain no notions of free will to consider him a danger to society.

Why is the conscious decision to do another person harm particularly blameworthy? Because what we do subsequent to conscious planning tends to most fully reflect the global properties of our minds—our beliefs, desires, goals, prejudices, etc. If, after weeks of deliberation, library research, and debate with your friends, you still decide to kill the king—well, then killing the king reflects the sort of person you really are. The point is not that you are the ultimate and independent cause of your actions; the point is that, for whatever reason, you have the mind of a regicide.

Certain criminals must be incarcerated to prevent them from harming other people. The moral justification for this is entirely straightforward: Everyone else will be better off this way. Dispensing with the illusion of free will allows us to focus on the things that matter—assessing risk, protecting innocent people, deterring crime, etc. However, certain moral intuitions begin to relax the moment we take a wider picture of causality into account. Once we recognize that even the most terrifying predators are, in a very real sense, unlucky to be who they are, the logic of hating (as opposed to fearing) them begins to unravel. Once again, even if you believe that every human being harbors an immortal soul, the picture does not change: Anyone born with the soul of a psychopath has been profoundly *unlucky*.

Why does the brain tumor in case 5 change our view of the situation so dramatically? One reason is that its influence has been visited upon a person who (we must assume) would not otherwise behave in this way. Both the tumor and its effects seem adventitious, and this makes the perpetrator appear to be purely a victim of biology. Of course, if we couldn't cure his condition, we would still need to lock him up to prevent him from committing further crimes, but we would not hate him or condemn him as evil. Here is one front on which I believe our moral intuitions must change: The more we understand the human mind in causal terms, the harder it becomes to draw a distinction between cases like 4 and 5.

The men and women on death row have some combination of bad genes, bad parents, bad environments, and bad ideas (and the innocent, of course, have supremely bad luck). Which of these quantities, exactly, were they responsible for? No human being is responsible for his genes or his upbringing, yet we have every reason to believe that these factors determine his character. Our system of justice should reflect an understanding that any of us could have been dealt a very different hand in life. In fact, it seems immoral not to recognize just how much luck is involved in morality itself.

To see how fully our moral intuitions must shift, consider what would happen if we discovered a cure for human evil. Imagine that every relevant change in the human brain could now be made cheaply, painlessly, and safely. In fact, the cure could be put directly into the food supply, like vitamin D. Evil would become nothing more than a nutritional deficiency.

If we imagine that a cure for evil exists, we can see that our retributive impulse is morally flawed. Consider, for instance, the prospect of *withholding* the cure for evil from a murderer as part of his punishment. Would this make any sense at all? What could it possibly mean to say that a person *deserves* to have this treatment withheld? What if the treatment was available prior to his crime? Would he still be responsible for his actions? It seems far more likely that those who had been aware of his case would be indicted for negligence. Would it make any sense to deny surgery to the man in case 5 as a *punishment* if we knew that the brain tumor was the actual cause of his violence? Of course not. The implications of this seem inescapable: The urge for retribution depends upon our not seeing the underlying causes of human behavior.

Despite our attachment to the notion of free will, most of us know that disorders of the brain can trump the best intentions of the mind. This shift in understanding represents progress toward a deeper, more consistent, and more compassionate view of our common humanity—and we should note that this is progress away from religious metaphysics. Few concepts have offered greater scope for human cruelty than the idea of an immortal soul that stands independent of all material influences, ranging from genes to economic systems. Within a religious framework, a belief in free will supports the notion of sin—which seems to justify not only harsh punishment in this life but *eternal* punishment in the next. And yet, ironically, one of the fears attending our progress in science is that a more complete understanding of ourselves will dehumanize us.

Viewing human beings as natural phenomena need not damage our system of criminal justice. If we could incarcerate earthquakes and hurricanes for their crimes, we would build prisons for them as well. We fight emerging epidemics—and even the occasional wild animal—without attributing free will to them. Clearly, we can respond intelligently to the threat posed by dangerous people without lying to ourselves about the ultimate origins of human behavior. We will still need a criminal justice system that attempts to accurately assess guilt and innocence along with the future risks that the guilty pose to society. But the logic of punishing people will come undone—unless we find that punishment is an essential component of deterrence or rehabilitation.

It must be admitted, however, that the issue of retribution is a tricky one. In a fascinating article in *The New Yorker*,[21] Jared Diamond writes of the high price we sometimes pay when our desire for vengeance goes unfulfilled. He compares the experiences of two people: his friend Daniel, a New Guinea highlander who avenged the death of a paternal uncle; and his late father-in-law, who had the opportunity to kill the man who murdered his entire family during the Holocaust but opted instead to turn him over to the police. (After spending only a year in jail, the killer was released.) The consequences of taking revenge in the first case and forgoing it in the second could not have been starker. While there is much to be said against the vendetta culture of the New Guinea highlands, Daniel's revenge brought him exquisite relief. Whereas Diamond's father-in-law spent the last 60 years of his life "tormented by regret and guilt." Clearly, vengeance answers to a powerful psychological need in many of us.

We are deeply disposed to perceive people as the authors of their actions, to hold them responsible for the wrongs they do us, and to feel that these transgressions must be punished. Often, the only punishment that seems appropriate is for the perpetrator of a crime to suffer or forfeit his life. It remains to be seen how a scientifically informed system of justice might steward these impulses. Clearly, a full account of the causes of human behavior should attenuate our natural response to injustice, at least to some degree. I doubt, for instance, that Diamond's father-in-law would have suffered the same anguish if his family had been trampled by an elephant or laid low by cholera. Similarly, we can assume that his regret would have been significantly eased if he had learned that his family's killer had lived a flawlessly moral life until a virus began ravaging his medial prefrontal cortex.

However, it may be that a sham form of retribution would still be moral—even necessary—if it led people to behave better than they otherwise would. Whether it is useful to emphasize the punishment of certain criminals—rather than their containment or rehabilitation—is a question for social and psychological science. But it seems clear that a desire for retribution, arising from the idea that each person is the free author of his thoughts and actions, rests on a cognitive and emotional illusion—and perpetuates a moral one.

One way of viewing the connection between free will and moral responsibility is to note that we generally attribute these qualities to people only with respect to actions that punishment might deter.[22] I cannot hold you responsible for behaviors that you could not possibly control. If we made sneezing illegal, for instance, some number of people would break the law no matter how grave the consequences. A behavior like kidnapping, however, seems to require conscious deliberation and sustained effort at every turn—hence it should admit of deterrence. If the threat of punishment could cause you to stop doing what you are doing, your behavior falls squarely within conventional notions of free will and moral responsibility.

It may be true that strict punishment—rather than mere containment or rehabilitation—is necessary to prevent certain crimes. But punishing people purely for pragmatic reasons would be very different from the approach that we currently take. Of course, if punishing bacteria and viruses would prevent the emergence of pandemic diseases, we would mete out justice to them as well.

A wide variety of human behaviors can be modified by punishments and incentives—and attributing responsibility to people in these contexts is quite natural. It may even be unavoidable as a matter of convention. As the psychologist Daniel Wegner points out, the idea of free will can be a tool for understanding human behavior. To say that someone freely chose to squander his life's savings at the poker table is to say that he had every opportunity to do otherwise and that nothing about what he did was inadvertent. He played poker not by accident or while in the grip of delusion but because he wanted to, intended to, and decided to, moment after moment. For most purposes, it makes sense to ignore the deep causes of desires and intentions—genes, synaptic potentials, etc.—and focus instead on the conventional outlines of the person. We do this when thinking about our own choices and behaviors—because it's the easiest way to organize our thoughts and actions. Why did I order beer instead of wine? Because I prefer beer. Why do I prefer it? I don't know, but I generally have no need to ask. Knowing that I like beer more than wine is all I need to know to function in a restaurant. Whatever the reason, I prefer one taste to the other. Is there freedom in this? None whatsoever. Would I magically reclaim my freedom if I decided to spite my preference and order wine instead? No, because the roots of this intention would be as obscure as the preference itself.

Politics

For better or worse, dispelling the illusion of free will has political implications—because liberals and conservatives are not equally in thrall to it. Liberals tend to understand that a person can be lucky or unlucky in all matters relevant to his success. Conservatives, however, often make a religious fetish of individualism. Many seem to have absolutely no awareness of how fortunate one must be to succeed at anything in life, no matter how hard one works. One must be lucky to be *able to work*. One must be lucky to be intelligent, physically healthy, and not bankrupted in middle age by the illness of a spouse.

Consider the biography of any "self-made" man, and you will find that his success was entirely dependent on background conditions that he did not make and of which he was merely the beneficiary. There is not a person on earth who chose his genome, or the country of his birth, or the political and economic conditions that prevailed at moments crucial to his progress. And yet, living in America, one gets the distinct sense that if certain conservatives were asked why they weren't born with club feet or orphaned before the age of five, they would not hesitate to take credit for these accomplishments.

Even if you have struggled to make the most of what nature gave you, you must still admit that your ability and inclination to struggle is part of your inheritance. How much credit does a person deserve for not being lazy? None at all. Laziness, like diligence, is a neurological condition. Of course, conservatives are right to think that we must encourage people to work to the best of their abilities and discourage free riders wherever we can. And it is wise to hold people responsible for their actions when doing so influences their behavior and brings benefit to society. But this does not mean that we must be taken in by the illusion of free will. We need only acknowledge that efforts matter and that people can change. We do not change ourselves, precisely—because we have only ourselves with which to do the changing—but we continually influence, and are influenced by, the world around us and the world within us. It may seem paradoxical to hold people responsible for what happens in their corner of the universe, but once we break the spell of free will, we can do this precisely to the degree that it is useful. Where people can change, we can demand that they do so. Where change is impossible, or unresponsive to demands, we can chart some other course. In improving ourselves and society, we are working directly with the forces of nature, for there is nothing but nature itself to work with.

Conclusion

It is generally argued that our experience of free will presents a compelling mystery: On the one hand, we can't make sense of it in scientific terms; on the other, we *feel* that we are the authors of our own thoughts and actions. However, I think that this mystery is itself a symptom of our confusion. It is not that free will is simply an illusion—our experience is not merely delivering a distorted view of reality. Rather, we are mistaken about our experience. Not only are we not as free as we think we are—we do not feel as free as we think we do. Our sense of our own freedom results from our not paying close attention to what it is like to be us. The moment we pay attention, it is possible to see that free will is nowhere to be found, and our experience is perfectly compatible with this truth. Thoughts and intentions simply arise in the mind. What else could they do? The truth about us is stranger than many suppose: *The illusion of free will is itself an illusion.*

The problem is not merely that free will makes no

sense objectively (i.e., when our thoughts and actions are viewed from a third-person point of view); it makes no sense subjectively either. It is quite possible to notice this through introspection. In fact, I will now perform an experiment in free will for all to see: I will write anything I want for the rest of this book. Whatever I write will, of course, be something I choose to write. No one is compelling me to do this. No one has assigned me a topic or demanded that I use certain words. I can be ungrammatical if I pleased. And if I want to put a rabbit in this sentence, I am free to do so.

But paying attention to my stream of consciousness reveals that this notion of freedom does not reach very deep. Where did this rabbit come from? Why didn't I put an elephant in that sentence? I do not know. I am free to change "rabbit" to "elephant," of course. But if I did this, how could I explain it? It is impossible for me to know the cause of either choice. Either is compatible with my being compelled by the laws of nature or buffeted by the winds of chance; but neither looks, or feels, like freedom. Rabbit or elephant? Am I free to decide that "elephant" is the better word *when I just do not feel that it is the better word?* Am I free to change my mind? Of course not. It can only change *me*.

What brings my deliberations on these matters to a close? This book must end sometime—and now I want to get something to eat. Am I free to resist this feeling? Well, yes, in the sense that no one is going to force me at gunpoint to eat—but I am hungry. Can I resist this feeling a moment longer? Yes, of course—and for an indeterminate number of moments thereafter. But I don't know why I make the effort in this instance and not in others. And why do my efforts cease precisely when they do? Now I feel that it really is time for me to leave. I'm hungry, yes, but it also seems that I've made my point. In fact, I can't think of anything else to say on the subject. And where is the freedom in *that*?

ACKNOWLEDGMENTS

I would like to thank my wife and editor, Annaka Harris, for her contributions to *Free Will.* As is always the case, her insights and recommendations greatly improved the book. I don't know how she manages to raise our daughter, work on her own projects, and still have time to edit my books—but she does. I am extremely lucky and grateful to have her in my corner.

Jerry Coyne, Galen Strawson, and my mother also read an early draft of the manuscript and provided very helpful comments.

NOTES

1. Recent advances in experimental psychology and neuroimaging have allowed us to study the boundary between conscious and unconscious mental processes with increasing precision. We now know that at least two systems in the brain—often referred to as "dual processes"—govern human cognition, emotion, and behavior. One is evolutionarily older, unconscious, slow to learn, and quick to respond; the other evolved more recently and is conscious, quick to learn, and slow to respond. The phenomenon of priming, in which subliminally presented stimuli influence a person's thoughts and emotions, exposes the first of these systems and reveals the reality of complex mental processes at work beneath the level of conscious awareness. People can be primed in a wide variety of ways, and these unconscious influences reliably alter their goals and subsequent behavior (H. Aarts, R. Custers, & H. Marien, 2008. Preparing and motivating behavior outside of awareness. *Science* 319[5780]: 1639; R. Custers & H. Aarts, 2010. The unconscious will: How the pursuit of

goals operates outside of conscious awareness. *Science* 329 [5987]: 47–50).

The experimental technique of "backward masking" has been at the center of much of this work: If one presents subjects with a brief visual stimulus (around 30 milliseconds), they can consciously perceive it; but they can no longer do so if this same stimulus is immediately followed by a dissimilar pattern (the "mask"). This technique allows for words and images to be delivered to the mind subliminally. Interestingly, the threshold for the conscious recognition of emotional words is lower than for neutral words, which suggests that semantic processing occurs prior to consciousness (R. Gaillard, A. Del Cul, L. Naccache, F. Vinckier, L. Cohen, & S. Dehaene, 2006. Nonconscious semantic processing of emotional words modulates conscious access. *Proc. Natl. Acad. Sci.* USA 103[19]: 7524–7529).

Recent neuroimaging experiments have offered further evidence: Masked words engage areas associated with semantic processing (M. T. Diaz & G. McCarthy, 2007. Unconscious word processing engages a distributed network of brain regions. *J. Cogn. Neurosci.* 19[11]: 1768–1775; S. Dehaene, L. Naccache, L. Cohen, D. Le Bihan, J. F. Mangin, J. B. Poline, et al., 2001. Cerebral mechanisms of word masking and unconscious repetition priming. *Nat. Neurosci.* 4[7]: 752–758; S. Dehaene, L. Naccache, H. G. Le Clec, E. Koechlin,

M. Mueller, G. Dehaene-Lambertz, et al., 1998. Imaging unconscious semantic priming. *Nature* 395[6702]: 597–600); subliminally promised rewards alter activity in the brain's reward regions and influence subsequent behavior (M. Pessiglione, L. Schmidt, B. Draganski, R. Kalisch, H. Lau, R. J. Dolan, et al., 2007. How the brain translates money into force: A neuroimaging study of subliminal motivation. *Science* 316[5826]: 904–906); and masked fearful faces and emotional words drive activity in the amygdala, the hub of emotional processing in the limbic system (P. J. Whalen, S. L. Rauch, N. L. Etcoff, S. C. McInerney, M. B. Lee, & M. A. Jenike, 1998. Masked presentations of emotional facial expressions modulate amygdala activity without explicit knowledge. *J. Neurosci.* 18[1]: 411–418; L. Naccache, R. Gaillard, C. Adam, D. Hasboun, S. Clemenceau, M. Baulac, et al., 2005. A direct intracranial record of emotions evoked by subliminal words. *Proc. Natl. Acad. Sci.* USA 102[21]: 7713–7717).

The subliminal presentation of stimuli poses some conceptual problems, however. As Daniel Dennett points out, it can be difficult (or impossible) to distinguish what was experienced and then forgotten from what was never experienced in the first place—see his insightful discussion of Orwellian vs. Stalinesque processes in cognition (D. C. Dennett, 1991. *Consciousness explained*. Boston: Little, Brown and Co., pp. 116–125).

This ambiguity is largely attributable to the fact that the contents of consciousness must be integrated over time—around 100 to 200 milliseconds (F. Crick & C. Koch, 2003. A framework for consciousness. *Nat. Neurosci.* 6[2]: 119–126). This period of integration allows the sensation of touching an object and the associated visual perception of doing so, which arrive at the cortex at different times, to be experienced as though they were simultaneous. Consciousness, therefore, is dependent upon what is generally known as "working memory." Many neuroscientists have made this same point (J. M. Fuster, 2003. *Cortex and mind: Unifying cognition*. Oxford: Oxford University Press; P. Thagard & B. Aubie, 2008. Emotional consciousness: A neural model of how cognitive appraisal and somatic perception interact to produce qualitative experience. *Conscious. Cogn.* 17(3): 811–834; B. J. Baars & S. Franklin, 2003. How conscious experience and working memory interact. *Trends Cogn. Sci.* 7(4): 166–172). The principle is somewhat more loosely captured by Gerald Edelman's notion of consciousness as "the remembered present" (G. M. Edelman, 1989. *The remembered present: A biological theory of consciousness*. New York: Basic Books).

2. B. Libet, C. A. Gleason, E. W. Wright, & D. K. Pearl, 1983. Time of conscious intention to act in relation to onset of cerebral activity (readiness-potential): The unconscious initiation of a freely voluntary act, *Brain* 106 (Pt 3):

623–642; B. Libet, 1985. Unconscious cerebral initiative and the role of conscious will in voluntary action. *Behav. Brain Sci.* 8: 529–566. Another lab has since found that a person's judgment of when he intended to move can be shifted in time by giving him delayed sensory feedback of his actual movements. This suggests that such judgments are retrospective estimates based on the apparent time of movement and not based on an actual awareness of the neural activity that causes the movement (W. P. Banks & E. A. Isham, 2009). We infer rather than perceive the moment we decided to act. (*Psychological Science*, 20: 17–21).

However, Libet and others have speculated that the concept of free will might yet be saved: Perhaps the conscious mind is free to "veto," rather than initiate, complex action. This suggestion has always seemed absurd on its face—for surely the neural events that inhibit a planned action arise unconsciously as well.

3. J. D. Haynes, 2011. Decoding and predicting intentions. *Ann. NY Acad. Sci.* 1224(1): 9–21.
4. I. Fried, R. Mukamel, & G. Kreiman, 2011. Internally generated preactivation of single neurons in human medial frontal cortex predicts volition. *Neuron*, 69: 548–562; P. Haggard, 2011. Decision time for free will. *Neuron*, 69: 404–406.
5. The neuroscientists Joshua Greene and Jonathan Cohen make a similar point:

> Most people's view of the mind is implicitly dualist and libertarian and not materialist and compatibilist. . . . That is, it requires the rejection of determinism and an implicit commitment to some kind of magical mental causation . . . contrary to legal and philosophical orthodoxy, determinism really does threaten free will and responsibility as we intuitively understand them (J. Greene & J. Cohen, 2004. For the law, neuroscience changes nothing and everything. *Philos. Trans. R. Soc. Lond. B Biol. Sci.* 359[1451]: 1775–1785).

6. For a good survey of compatibilist thought, see http://plato.stanford.edu/entries/compatibilism/. See also G. Watson, ed., 2003. *Free will* (second edition). Oxford: Oxford University Press.
7. D. C. Dennett, 2003. *Freedom evolves*. New York: Penguin.
8. Tom Clark, personal communication.
9. Daniel Dennett, personal communication.
10. Galen Strawson (personal communication) has pointed out that even if one agrees with Dennett here, the ordinary notion of moral responsibility is still deeply problematic for the reasons already given.
11. In his book *Consciousness Explained*, Daniel Dennett describes an unpublished experiment in which the neurosurgeon W. Grey Walter directly connected the motor cortices of his patients to a slide projector. Asked to advance the slides at their leisure, the subjects were said to have

felt that the projector was reading their minds. Unfortunately, there is some uncertainty as to whether the experiment was ever performed.

12. D. Wegner, 2002. *The illusion of conscious will.* Cambridge, MA: Bradford Books/MIT Press.
13. L. Silver, 2006. *Challenging nature: The clash of science and spirituality at the new frontiers of life.* New York: Ecco, p. 50.
14. For a recent discussion of the role of consciousness in human psychology, see R. F. Baumeister, E. J. Masicampo, & K. D. Vohs, 2011. Do conscious thoughts cause behavior? *Annual Review of Psychology*, 62: 331–361.
15. Again, as Galen Strawson points out (personal communication), even if we granted that you are the whole of your mind (conscious and unconscious), you still cannot ultimately be held responsible for its character.
16. Einstein (following Schopenhauer) once made the same point:

> Honestly, I cannot understand what people mean when they talk about the freedom of the human will. I have a feeling, for instance, that I will something or other; but what relation this has with freedom I cannot understand at all. I feel that I will to light my pipe and I do it; but how can I connect this up with the idea of freedom? What is behind the act of willing to light the pipe? Another act of willing? Schopenhauer once said: Der Mensch kann was er will; er kann aber nicht wollen

was er will (Man can do what he will but he cannot will what he wills). (M. Planck, 1932. *Where is science going?* New York: W. W. Norton & Company, p. 201.)

17. As Jerry Coyne points out (personal communication), this notion of counterfactual freedom is also scientifically untestable. What evidence could possibly be put forward to show that one could have acted differently in the past?
18. http://opinionator.blogs.nytimes.com/2011/11/13/is-neuroscience-the-death-of-free-will/.
19. K. D. Vohs & J. W. Schooler, 2008. The value of believing in free will: Encouraging a belief in determinism increases cheating. *Psychological Science*, 19(1): 49–54.
20. R. F. Baumeister, E. J. Masicampo, & C. N. DeWall, 2009. Prosocial benefits of feeling free: Disbelief in free will increases aggression and reduces helpfulness. *Personality and Social Psychology Bulletin*, 35: 260–268.
21. J. Diamond, 2008. Vengeance is ours. *The New Yorker*, April 21, 2001, pp. 74–87.
22. Steven Pinker, personal communication.

INDEX

ABOUT THE AUTHOR

Sam Harris is the author of the bestselling books *The End of Faith, Letter to a Christian Nation, The Moral Landscape,* and *Lying. The End of Faith* won the 2005 PEN Award for Nonfiction. His writing has been published in more than 15 languages. Dr. Harris and his work have been discussed in *The New York Times, Scientific American, Nature, Rolling Stone*, *Newsweek, Time*, and many other publications. His writing has appeared in *The New York Times*, the *Los Angeles Times, The Times* (London), the *Boston Globe, The Atlantic, Newsweek*, the *Annals of Neurology*, and elsewhere. Dr. Harris is a cofounder and the CEO of Project Reason, a nonprofit foundation devoted to spreading scientific knowledge and secular values in society. He received a degree in philosophy from Stanford University and a PhD in neuroscience from UCLA. Please visit his website at www.samharris.org.